Performance Evaluation of Vision Algorithms on FPGA

Mahendra Gunathilaka Samarawickrama

DISSERTATION.COM

Boca Raton

Performance Evaluation of Vision Algorithms on FPGA

Dissertation.com
Boca Raton, Florida
USA • 2010

ISBN-10: 1-59942-373-1
ISBN-13: 978-1-59942-373-9

Abstract

The modern FPGAs enable system designers to develop high-performance computing (HPC) applications with large amount of parallelism. Real-time image processing is such a requirement that demands much more processing power than a conventional processor can deliver. In this research, we implemented software and hardware based architectures on FPGA to achieve real-time image processing. Furthermore, we benchmark and compare our implemented architectures with existing architectures. The operational structures of those systems consist of on-chip processors or custom vision coprocessors implemented in a parallel manner with efficient memory and bus architectures. The performance properties such as the accuracy, throughput and efficiency are measured and presented.

According to results, FPGA implementations are faster than the DSP and GPP implementations for algorithms which can exploit a large amount of parallelism. Our image pre-processing architecture is nearly two times faster than the optimized software implementation on an Intel Core 2 Duo GPP. However, because of the higher clock frequency of DSPs/GPPs, the processing speed for sequential computations on on-chip processors in FPGAs is slower than on DSPs/GPPs. These on-chip processors are well suited for multi-processor systems for software level parallelism. Our quad-Microblaze architecture achieved 75-80% performance improvement compared to its single Microblaze counterpart. Moreover, the quad-Microblaze design is faster than the single-powerPC implementation on FPFA. Therefore, multi-processor architecture with customised coprocessors are effective for implementing custom parallel architecture to achieve real time image processing.

To my loving wife Prasadi, parents and teachers for giving me constant support and motivation.

Acknowledgment

I wish to thank my supervisors Dr. Ajith Pasqual and Dr. Ranga Rodrigo for their support and encouragement during this research. Their insight, guidance, feedback and especially the constructive criticisms contributed enormously to the production of this thesis.

I am grateful to Dr. E.C. Kulasekere, the coordinator of this research and Dr. Chathura De Silva, the chairman of the progress review committies and Prof. (Mrs.) I.J. Dayawansa, the postgraduate research advisor for their feedback, kind advice and invaluable suggestions given.

I am deeply indebted to other academics and administrators who have provided helpful advice and knowledge during this research.

I also wish to extend my gratitude to Zone24×7 (Pvt) Ltd. for providing laboratory facilities.

I acknowledge the financial support given by the *University of Moratuwa Senate Research Committee grant SRC-297*, to enable me conduct the masters program at University of Moratuwa.

Finally, I am thankful to my parents, family and friends for their care, commitment and support they extended to me during this research program.

MAHENDRA G. SAMARAWICKRAMA

July 2010

Contents

List of Figures

List of Tables

Abbreviations

Following abbreviations or acronyms have been used in this thesis.

Abbreviations/acronyms	Meaning
ADDR	Address: Memory location for read/write data
BRAM	Block RAM
CLK	Clock
CMP	Chip Multiprocessor
DDR	Double Data Rate
DIN	Data Input: Data written into memory
DOUT	Data Output: Synchronous output of the memory
DSP	Digital Signal Processor
EDK	Embedded Development Kit
EN	Enable: Enables access to memory
EEPROM	Electrically Erasable Programmable ROM
FPGA	Field-Programmable Gate Array
GPP	General Purpose Processor
HDL	Hardware Description Language
HLL	High-Level Language
HLS	High-Level Synthesis
LMB	Local Memory Bus
LUT	Lookup Table
ROM	Read-Only Memory
PIF	Performance Improve Factor
PLB	Processor Local Bus
SoC	System on Chip
WE	Write Enable: Allows data transfer into memory
XCL	Xilinx Cache Link
XPS	Xilinx Platform Studio

Nomenclature

Following symbols or notations have been used in this thesis.

Notation	Meaning
T_{nbhd}	Time to read neighborhood pixels around the first pixel of the image
N_{mask}	Kernel dimension
f_{clk}	Clock frequency
T_{img}	Total time needs to process all the pixels of the image
M_{img}	Number of pixels per image
T_{SM}	Time to execute in single-microblaze architecture
T_{QM}	Time to execute in quad-processor-microblaze architecture

CHAPTER 1

Introduction

1.1 Background

Real-time image processing is a primary requirement for applications such as navigation and tracking. Real-time image processing requires processing an image at video rate, i.e., at least 30 images per second. There are several technologies that are available to achieve this goal. For example, DSPs, FPGAs and GPPs can be mentioned. However, each technology has advantages and disadvantages when compared to the other. In fact, their usability depends on the requirements related to power consumption, cost, flexibility and design cycle time. Regarding the power consumption, DSPs and FPGAs are more suitable than GPPs. When considering the development cost of DSPs, it is very expensive and suitable only where there is mass scale production. Moreover, in DSPs, time-to-market is high and time-in-market is low compared to FPGAs. Nevertheless, available DSPs are generally cheaper and faster in performance than FPGAs. Regarding the GPPs, when they are implemented with software libraries with guaranteed high efficiency in machine level instructions handling, modern multi-cores GPPs run very fast. As an example, we can show the compatibility of Intel Core i7 and OpenCV library. In the case of design flexibility, FPGAs are more suitable with their parallel and reconfigurable architecture. However, depending on the designing approach, FPGA-based algorithm implementations are often complex and tedious compared to DSPs and GPPs.

1

Therefore, modern real-time vision systems are not relying upon a single technology, but more converge technologies to get the advantages of each technology. Fig. 1.1 shows, Texas machine vision solution [1] which is a good example for technology convergence. It is composed of FPGA, DSP and interface to communicate with the GPP. However, design flexibility in such systems is limited because of their heterogeneous nature.

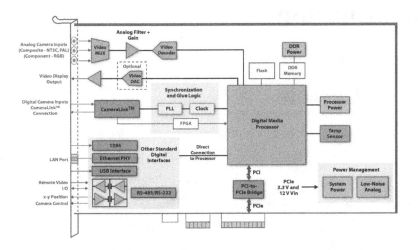

Figure 1.1: Texas machine vision solution [1] is a good example for technology convergence. It is composed of FPGA, DSP and interface to communicate with GPP.

By focusing on FPGAs as a flexible and reconfigurable solution, in recent years (2008-2010), FPGA devices are developed with significantly higher amount of internal memory and logic resources with much higher bandwidth [2]. Fig. 1.2 shows the growth of FPGA memory, resources and bandwidth in the past decade. This growth gives system designers a good opportunity to design vision coprocessors which consume large amount of hardware resources.

Moreover, it can be seen that (Fig. 1.3), high-performance computing (HPC) applications' demands have outpaced the conventional processors' performance. Therefore, as a solution to the real-time image processing, it is possible to make hardware acceleration with application-specific coprocessors. Due to the right

2

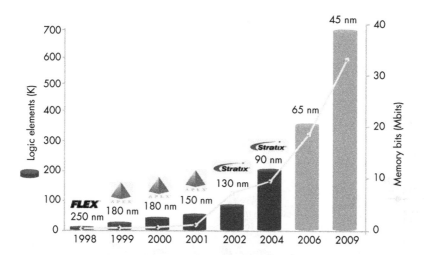

Figure 1.2: The growth of FPGA memory, logic resources and bandwidth [2]. This growth gives system designers good opportunity to design vision coprocessors which consume large amount of hardware resources.

combination of price, performance, and ease-of use, along with significant power savings, FPGA is a good option for coprocessor designing.

FPGA devices have not only grown in terms of resources and speed but also in the amount of embedded processors within their fabric. These embedded processors together with coprocessors are now capable of designing custom computers to achieve common tasks.

FPGAs offer many performance and implementation benefits for executing image processing applications [3]. The main advantage of FPGA-based design is the flexibility to exploit the inherently parallel nature of many image processing problems. For example, many vision algorithms require the repeated application of the same local operation, such as convolution, to every region in an image. In a serial processor this can be quite time consuming. However, in an FPGA, multiple convolutions can take place simultaneously.

There are many FPGA-based vision architectures that have been implemented with various types of resources and strategies. Huitzil and Estrada [4] proposed a design where FPGA device is incorporated with two external memory banks.

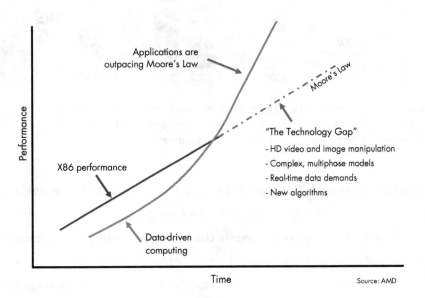

Figure 1.3: The technology gap between demand and performance of a real time system [2]. High-performance computing (HPC) applications are now demanding more than processors alone can deliver, creating a technology gap between demand and performance.

It can process the SUSAN algorithm for high-speed edge detection on 512×512 images with 50 MHz clock up to 120 images per second. An efficient address generator module with register-intensive hardware buffers significantly reduce the frequent real-time memory access, hence speed up the algorithm at the expense of high resource utilization.

FPGAs have also been used to process computationally complex algorithms fast. Wei *et al.* [5] implemented DDR memory based architecture for optical-flow algorithms with the help of Xilinx EDK tools. It can process 640×480 16-bit YUV images through an Ethernet interface up to 64 images per second. By implementing a PowerPC, rather than implementing a coprocessor to make calculations related to optical flow algorithm, they could handle floating points easily. It is a good tradeoff between accuracy and processing speed.

Modern FPGA/DSP systems have various types of hierarchical memories and caches that differ significantly in their sizes, latency times, and data bandwidths. Daniel Baumgartner *et al.* [6] showed that the memory selection, configuration and data access pattern have a significant influence on the achievable speed. From their results, Fig. 1.4 compares execution times of Gaussian pyramid function under different memory configurations on a C64x DSP. The results clearly show that memory access overhead critically slows down the algorithm execution speed.

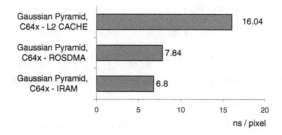

Figure 1.4: Execution times of Gaussian pyramid function under different memory configurations on a C64x DSP: IRAM (Internal RAM), ERAM (External RAM) + ROSDMA (Resource Optimized Slicing Direct Memory Access), ERAM + L2-Cache [6].

To improve the performance of FPGA-based implementations, pipeline insertion is one of the main strategies. Teifel and Manohar [7] showed that pipelines improve overall clock frequency and speed up the FPGA-based systems. Fig. 1.5 shows the characteristics of the improvement of maximum clock frequency against the number of pipeline stages. However, the speed improvement is not uniform. When more pipeline stages are added in the design the latency increases and the maximum clock frequency tends to decrease.

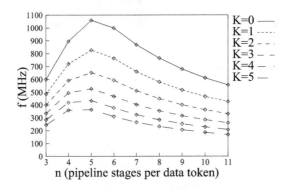

Figure 1.5: Maximum operating frequency curves for one token in a linear pipeline of n stages. K is the number of routing switches between every pipeline stage ($K=0$ is the "custom" case when there are no switches between stages) [7].

FPGA-based embedded processing requirements are growing at a rapid pace and system architects are not limited to coprocessors and are turning towards hybrid multiprocessor designs [8]. The Xilinx Platform Studio (XPS) and Embedded Development Kit (EDK) is a comprehensive solution for designing embedded programmable systems. These tools make it easy to design powerful chip multi-processing (CMP) systems.

Implementation of an algorithm on the FPGA requires building and utilising FPGA-specific hardware such as fast multipliers and BRAM, on an open architecture platform. This is fundamentally different to the design of software for the fixed architectures of conventional processors. Although software techniques may help to define the image processing algorithm and facilitate its programming, they will provide little guidance on how to manage hardware resources on the FPGAs.

Gribbon *et al.* [9] are actively working for visual programming language which can incorporate hardware design patterns for FPGA-based image processing.

Moreover, generating hardware by translating High-Level Language (HLL) to Hardware-Description Language (HDL) using application mappers, is developed as a modern trend. In other words, this method can likely increase performance and productivity while minimizing unnecessary development time and effort. Curreri *et al.* [10] in their survey state that the development time using VHDL, including the researchers algorithm-acquaintance period, was roughly three times longer than using AccelDSP. However, the HDL version has higher achievable core frequency because the automated procedure of translating HLL to HDL creates overhead that increases critical-path delays [10]. The resulting clock frequency, pipeline length, and selected resource utilization of the HLL and HDL implementations are shown in Table 3.5.

Table 1.1: HLL vs. HDL comparisons in core design [10]

Core Design	Design 1		Design 2	
	HLL	HDL	HLL	HDL
Core Freq. (MHz)	140	180	150	200
Pipeline Len. (cycle)	31	34	31	34
MULT18×18s (%)	5	5	5	5
Slices (%)	18	17	20	17

By analyzing above design strategies and concepts, we can identify several challenges, which are related to FPGA-based image processing.

1.2   Design Challenges

Complete development of an image processing application for an FPGA-based system involves four stages: (1) problem specification, (2) algorithm development, (3) architecture selection, and (4) implementation. Considering the algorithm development and implementation, simply mapping of a software implementation into hardware often falls short of the potential benefits offered by an FPGA solution as they do not tend to leverage concurrency. On the other hand, schematic entry and HDLs are often too low-level and designing image processing algorithms at

7

this level is complex, tedious and error prone [11]. Therefore, when implementing vision algorithms on FPGAs, we have to deal with development issues in each stage.

Memory configuration is an important design issue concerning architecture selection. Since there are several different memory resources on the FPGA systems (registers, BRAM and DDR-RAM), at the designing stage it is important to use them appropriately. For example, registers are used to store intermediate values in the algorithms, while BRAM are ideally suited for row buffering which is necessary for local filtering. Although on-chip BRAM is usually sufficient to buffer several rows of an image, the off-chip memory (DDR-RAM or SRAM) is often necessary for applications that require frame buffering.

Regarding the implementation of some image processing tasks, they are harder to implement and tend to be inefficient in custom hardware because they may not be running frequently. For example, histogram equalization runs once per frame as opposed to convolution operation which need to be run for each pixel in an image scan. This can leave hardware idle for long periods of time which is often an ineffective use of the limited logic resources. An alternative is to implement some of these tasks in a processor with software programmability in the same system. This is how, PowerPC or Microblazes based designing techniques are becoming increasingly important in high performance computing on FPGAs [8].

Therefor, implementation and integration of software and hardware on FPGA, can be seen as an effective solution in real-time vision applications. A finite state machine can be built on the FPGA fabric for performing intermediate-level tasks with low processing requirements. For more complicated processing, software/hardware processors (PowerPC or Microblazes) can be used. Implementing a processor based architecture provides several advantages. Instruction sets can be customized for efficiency and wiring delays are minimized. High development costs can be offset against software programmability; the same FPGA configuration can be used for several applications and a custom parallel computer can be built quickly [12].

CHAPTER 2

Design and Implementation

By referring designing issues described in the chapter 1, we made three approaches to implement different FPGA-based architectures to benchmark and compare their performance on real-time vision computing. In our first approach, we implemented image pre-processing architecture with gate-level parallelism. Then we implemented an image coprocessor using HLL synthesis tool and investigated its performance and advantages of HLL synthesis tools. Next, multi-processor architecture was implemented to benchmark software level parallelism. By realizing these architectures, we presented our comprehensive analysis, how modern FPGAs can be used in real-time image processing. In detailed description of architectures' implementations are described below as separate subsections.

2.1 Image Pre-Processing Architecture

A typical processing sequence of a vision system can be classified as a layered architecture (Fig. 2.1). The acquisition layer controls the sensor interface, pixel addressing, and passes source pixels to the pixel pre-processing layer, which in turn, performs corrections such as noise reduction and compensation [13].

For pixel pre-processing, it needs low amount of memory for buffering and can be implemented using on-chip block-memory on FPGAs. As an intermediate RAM structure, block RAMs [14] provide less complexity and speedy access (Fig. 2.4). Since the use of block RAMs do not affect the slices, they are well suited in terms

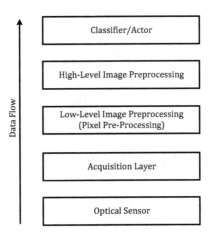

Figure 2.1: Layered architecture of a vision system

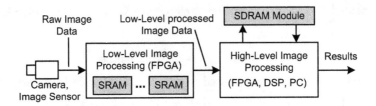

Figure 2.2: A typical vision system: The acquisition layer controls the sensor interface, pixel addressing, and passes source pixels to the pixel pre-processing layer, which, in turn, performs corrections such as noise reduction and compensation [13].

of FPGA resources.

Therefore, in our first approach to design a vision architecture, we developed an image pre-processing architecture which utilizes block RAMs for image buffering [15]. Furthermore, it consists of several organized modules (Fig. 2.3), which exploit parallelism with pipelining, loop splitting and independent operation execution.

In our architecture, we made a modular approach to achieve high flexibility in algorithm changes and modifications. The vision core is an application-specific co-processor where we implement image processing algorithms. The algorithm is initiated after the image or part of the image is stored in data store memory-bank. The results are stored in a separate memory-bank which is named as a

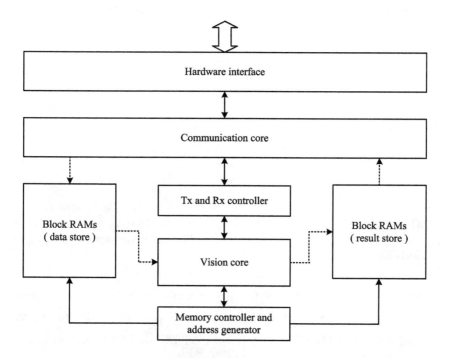

Figure 2.3: Image pre-processing architecture [15]: After the full frame is stored in the block-memory, the algorithm is initiated and results are stored in another block-memory. Dotted and continuous lines represent data and control busses, respectively.

result store. Since the memory controller and address generator modules are implemented apart from this module, the implementation of different algorithms is flexible and architecture independent.

In designing the memory-banks, we instantiated the basic RAM primitive (Fig. 2.4(a)) to achieve the maximum memory depth. However it limits the port-width of the memory to 1-bit and it required multi-level coupling. Fig. 2.4 and Fig. 2.5 shows, how we composed the memory-banks to achieve memory requirement.

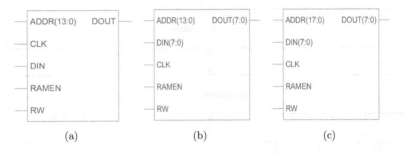

Figure 2.4: Implemented memory-bank schematics: (a) Basic schematic of RAMB16_S1 RAM primitive; (b) RAMB16_S1 coupled to store and access Bytes: this is maximum of 16KB; (c) Memory-bank which is implemented by coupling multiple 16KB RAMs.

Regarding the memory management, the address generator module creates the addressing sequence to access the image pixels that are needed to perform the vision algorithm. The address generator module is coupled with the memory-bank through the address bus ADDR(m:0). The width of the bus is determined by the required address depth. To support the address generator, there exists a memory controller module. It selects and enables/disables memory blocks while the algorithm is in progress. Once the address generator module generates the address of the data source or data store, this enables the required memory block and performs the read/write operation.

The communication core is mainly concerned with the transfer of image frames in and out through the FPGA interface. This interface can be Ethernet, PCI or a customized interface. Since the algorithm and memory access are performed at a much higher speed, design of the communication module is critically important

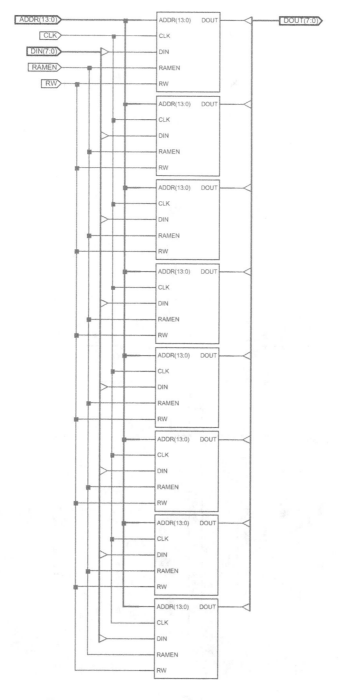

Figure 2.5: 16KB memory-block architecture

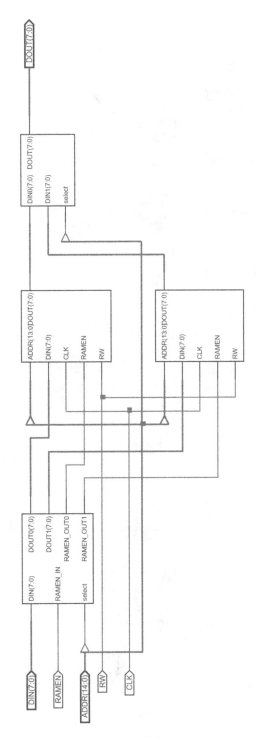

Figure 2.6: Memory-bank which is implemented with two 16KB memory blocks

14

to achieve higher throughput.

In neighborhood operations, we have mapped the two-dimensional image pixel array into a single dimensional memory array in the memory-bank. The neighborhood values of the image are distributed apart in the memory as shown in Fig. 2.7.

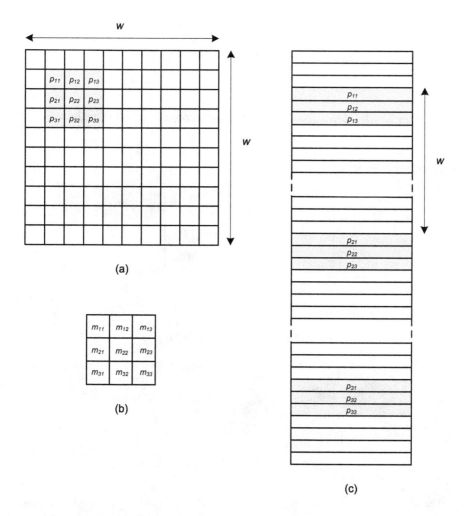

Figure 2.7: The model of mapping pixel array into memory array: (a) two dimensional image array; (b) filter mask; (c) single dimensional memory array.

In the convolution we took sum of products of the filter coefficients m_{xy} and

15

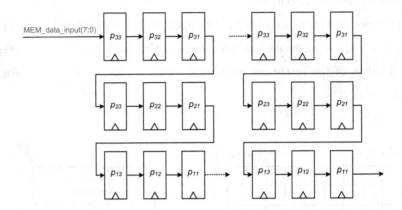

Figure 2.8: Internal data flow of the vision core for a 3×3 convolution

the corresponding image pixels p_{xy} in the area spanned by the filter mask.

$$M_{\text{mask}} = \begin{bmatrix} m_{11} & m_{12} & m_{13} \\ m_{21} & m_{22} & m_{23} \\ m_{31} & m_{32} & m_{33} \end{bmatrix}$$

$$
\begin{aligned}
R = |(p_{11}m_{11}) &+ (p_{12}m_{12}) + (p_{13}m_{13}) \\
&+ (p_{21}m_{21}) + (p_{22}m_{22}) + (p_{23}m_{23}) \\
&+ (p_{31}m_{31}) + (p_{32}m_{32}) + (p_{33}m_{33})| \quad (2.1)
\end{aligned}
$$

where,

$$
R = \begin{cases} \forall R \in [0, 255] & \text{if } M_{\text{Gaussian}} \\ \forall R \in [0, R_{\text{max}}] & \text{if } M_{\text{Laplacian}} \text{ or } M_{\text{Sobel}} \end{cases} \quad (2.2)
$$

The vision core, which is coupled to memory-banks is composed of 8 bit register pipeline (data pipeline). With each clock cycle the whole data pipeline is shifted by one pixel and a new pixel is read from MEM_data_input(7:0). Because of this data pipeline, one convolution operation is performed in a single clock cycle.

16

Refer to (Eq. 2.2), the result R of the Gaussian masking does not require rescaling since the resulting pixel intensity set is also a subset of the gray scale intensity set. On the other hand, the Laplacian operation results in a pixel intensity set which can exceed the 255 limit, necessitating implementation of rescaling. Therefore, we rescaled the image where gray levels in the set of $[0, R_{max}]$ are mapped into $[0, 255]$ set.

$$R_{scaled} = \frac{R}{\alpha} \tag{2.3}$$

$$\alpha = 2^{c\{\log_2(\frac{R_{max}}{256})\}} \tag{2.4}$$

$$c(z) = \begin{cases} 0 & z \leq 0 \\ \text{ceil}(z) & z > 0 \end{cases} \tag{2.5}$$

Since the rescaling algorithm uses the logarithm of base-two, all the dividers are of power-of-two. This is an added advantage, as at the hardware-level a division other than by a power-of-two needs subtraction and comparison which leads to an unacceptable time consumption.

With this architecture, we implemented several other low-level algorithms such as right-angle-corner detector and edge detector.

2.2 Image Coprocessor

It is hard to understand how a certain algorithm should be implemented on FPGA to obtain the most optimal realization. Implementations on FPGA which are written in HDL, leads to long realization time and this is a problem for rapid prototyping and testing high-level functionality in image processing. As an example, image pre-processing architecture which has been described above can be mentioned. As per the design flow, if it is possible to generate Verilog automatically from a high level description it would reduce the implementation time.

In this implementation Xilinx AccelDSP, a high-level synthesis (HLS) tool generating HDL from a high-level Matlab description has been evaluated. AccelDSP generated the Register Transfer Level (RTL) for the target FPGA device and applied optimizations as applicable and permissible by boundary conditions (i.e., performance requirements, logic resource availability and block RAM availability). Then Xilinx ISE, a RTL tool is used to transform the RTL implementation into a complete FPGA implementation in the form of a bitstream for programming a specific FPGA on a specific hardware platform with I/O and memory. The design flow of AccelDSP in conjunction with ISE is shown in Figure 2.9.

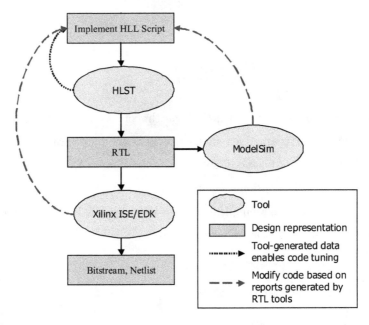

Figure 2.9: Design flow of AccelDSP high-level synthesis tool in conjunction with ISE RTL tool.

AccelDSP lets the user execute the script file inside the program and shows all plots, variables and output. The floating-point model is which must be verified by the designer using this output. A plot presenting the values and differences between the models are shown when running the AccelDSP function. It can be used to identify errors in the quantization. Errors in this model will propagate

18

through all later steps and exist in the final bitstream. It is also important to check that all important variables are observed since the output is used to verify the fixed-point model. After the program has verified that the fixed-point model of the algorithm corresponds to the intended function, AccelDSP generates a Register Transfer Level (RTL) model. A testbench for verifying the generated design is also created automatically (Fig. 2.10).

The testbench is then executed with ModelSIM and presents the results in a waveform table. The waveform is used to calculate the execution time and start-up cycles of the generated design. The testbench will also run a verification procedure that compares the output waveform of the RTL model with the reference data files. The result of this procedure will report pass or fail.

If the verification passes, the synthesis can be performed and gate-level netlist can be created. The netlist is then mapped to the hardware and generates the configuration bitstream which is used for programming the FPGA [16].

In our design, image data are stored in the block RAM when they are transferred into the FPGA through the communication core. Initially all the neighborhood pixels around the first pixel of the image corresponding to the convolution mask are read in from the block RAM to 8 bit registers. It will take T_{nbhd} time which is expressed by Eq. 2.6 where f_{clk} is the clock frequency to the coprocessor and N_{mask} is the kernel dimension (i.e., kernel size 3×3 $\Rightarrow N_{\mathrm{mask}} = 3$).

$$T_{\mathrm{nbhd}} = \frac{N^2{}_{\mathrm{mask}}}{f_{\mathrm{clk}}} \qquad (2.6)$$

After this initial memory access we can shift the convolution mask along the image where we need to access only the newly introduced N_{mask} number of pixels. This amount of time is taken for every other convolution. Therefore, total time needed to process all the pixels of the image, T_{img}, can be expressed by Eq. 2.7 where M_{img} is the number of pixels per image. This is a significant speed improvement compared to accessing all the neighborhood pixels in each time.

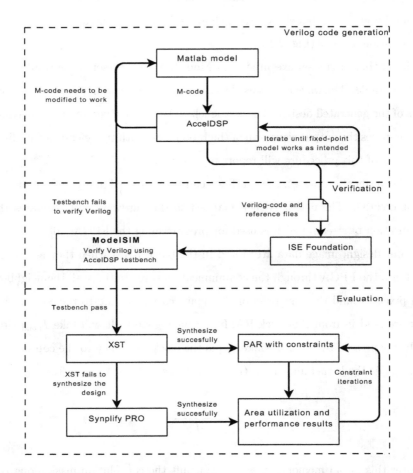

Figure 2.10: The Verilog generation workflow and evaluation. AccelDSP translate M-Code into Verilog which can be synthesized to create digital hardware. The translation includes automatic analysis of variables and generation of a fixed-point design suitable for hardware implementation.

$$T_{\text{img}} = T_{\text{nbhd}} + \frac{M_{\text{img}}N_{\text{mask}}}{f_{\text{clk}}} \tag{2.7}$$

$$T_{\text{img}} = \frac{1}{f_{\text{clk}}} \cdot (N^2{}_{\text{mask}} + M_{\text{img}}N_{\text{mask}}) \tag{2.8}$$

$$T_{\text{img}} \approx \frac{1}{f_{\text{clk}}} \cdot (M_{\text{img}}N_{\text{mask}}) \tag{2.9}$$

Our coprocessor has flexibility to extend the kernel size from 3×3 to 15×15. The variables can easily be coded in Verilog to meet the requirement. Therefore, this coprocessor is able to implement low-level operations without changing the architecture.

2.3   Standalone PowerPC Architecture

In image processing, there are some algorithms which are difficult to find a way to parallelism (i.e., complex branches (if-else), control loops and complex trigonometry). For example, to perform histogram equalization entire image pixels need to be analyzed sequentially. Since it is impossible to load the entire image into registers, this has little gain in hardware parallelism. Further, if there is no such gain on hardware parallelism in particular algorithm or stage, it may be a waste of resources going into hardware implementation. Moreover, implementing a slow hardware module can cause the entire system to slow down. Embedded processors on FPGAs (i.e., PowerPC and Microblaze) provide a reconfigurable solution to such situations and save FPGA resources while implementing them on the same FPGA.

In our standalone PowerPC design (Fig. 2.11), the processor has two ports for instruction fetch and data fetch. These ports are connected to Processor Local Bus (PLB). PowerPC has a dual-bus architecture which has another two ports for instruction and data fetch. These two ports are connected to the DDR2 RAM controller via PLB. Thus a dedicated bus exists for RAM access. Each hardware controller is assigned with an address space so PowerPC can access each hardware controller in a memory mapped fashion. This architecture is capable of operating at 341.35 MHz clock speed. Since this PowerPC processor is capable of achieving maximum of 400 MHz clock speed, our architecture is highly efficient.

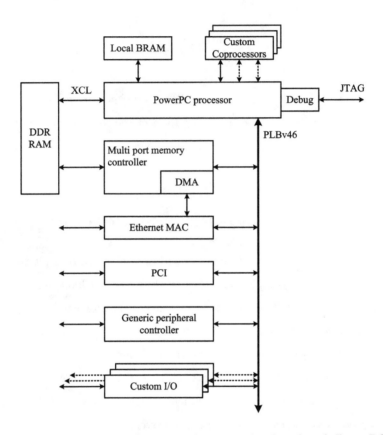

Figure 2.11: Bus interface of vision architecture developed with PowerPC-405

2.4 Single Microblaze Architecture

Single Microblaze architecture (Fig. 2.12) also has a similar architecture as the PowerPC platform. Since Microblaze is a soft core, this architecture operates with the maximum clock speed of 176.96 MHz. This is significantly slower than

PowerPC architecture. Therefore, speed performance what we can achieve with a single Microblaze architecture is not that satisfactory.

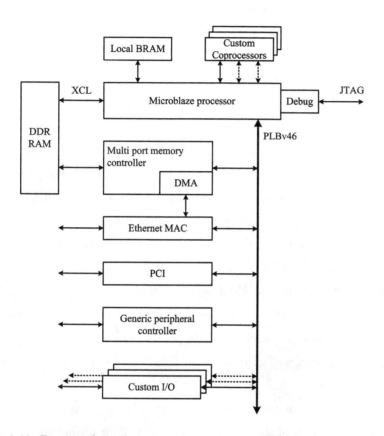

Figure 2.12: Bus interface of vision architecture developed with single Microblaze

2.5   Multiple-Microblaze Architecture

Embedded processors are no longer relied upon increasing clock speeds to increase processor performance. Heat dissipation and power consumption become increasingly problematic at high clock speeds (above 400 MHz) calling for other techniques to be considered [17]. One option is to leverage concurrency by building chips incorporating multiple processors. Our multiple microblaze architecture

(Fig. 2.13) is such a system which incoperates four microblaze processors to handle multi-frame processing in vision.

There are four Microblaze processors where each is provided with a separate BRAM to hold program code. LMB interface of each Microblaze processor is used to fetch program instructions and data from the BRAM via BRAM controller. Each processor is connected to a PLB which holds their respective hardware controllers. A four port MUTEX is used to achieve synchronization between processors. As these processor buses are independent, the hardware IO controllers on the primary processor can not be accessed by other processors. To overcome this problem a bus bridge is used which brings a portion of address space of the primary processor into visibility of the three secondary processors. At this point the design is technically a working platform. But if data is fetched from DDR2 RAM via PLB, it would be extreamly slow since burst data transfer is not pipelined or cached. To use the special data transfer features supported by the DDR2 RAM, we implemented Xilinx Cache Link (XCL) by creating eight XCL ports on the DDR2 RAM controller. By that we could improve the memory access performance by 40 times. To transfer data, each processor grabs the mutex lock (Fig. 2.14) and accesses the communication hardware controller attached to primary processor via shared address space.

There are several benefits of this architecture. (1) Programs can be coded in C language. Since most of the high-level vision algorithms are incorporated with complex trigonometric functions, it is effective and easier to implement them in C. (2) With the four processors, programs can be executed in multiple threads. Mutex is used in concurrent programming to avoid the simultaneous use of a common resources (memory and bus). (3) Dynamic memory allocation is possible. This is helpful when we need to implement linked list and dynamic arrays.

To measure the algorithm speed and benchmark the architecture we followed the following steps. First, we send and receive an image without performing any algorithm on that image. Then, we send and receive the image while performing histogram-equalization algorithm on it. The time difference of two steps is purely the time taken to perform histogram-equalization. Since this time differ-

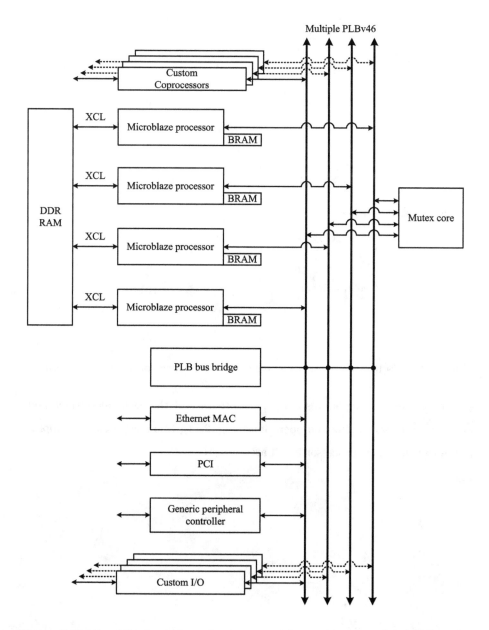

Figure 2.13: Bus interface of quad-processor vision architecture developed with four Microblazes

```
#include"hist.h"
#include"mymutex.h"

volatile XMutex mutex;

void main ()
{
   microblaze_init_icache_range (0, XPAR_MICROBLAZE_CACHE_BYTE_SIZE);

   microblaze_enable_icache ();
   microblaze_init_dcache_range (0, XPAR_MICROBLAZE_DCACHE_BYTE_SIZE);

   microblaze_enable_dcache ();          //REMARKABLE 40 TIMES SPEEDUP

   microblaze_disable_interrupts();
   microblaze_disable_exceptions();

   init_lock(&mutex);

   XMutex_Lock(&mutex,0);
   UartInit();
   XMutex_Lock(&mutex,0);

   while(1)
   {
      histeq(640,480,&mutex);
   }

}
```

Figure 2.14: Software operation-sequence of quad-processor vision architecture

ence expresses memory access delays, performance of the cache algorithms and
multiprocessor scheduling and operations (Fig. 2.14), this methodology is appro-
priate to benchmark processor-based architectures.

CHAPTER 3

Results

We coupled a computer to the FPGA to upload image sequence and to download results. This is feasible through any type of communication interface which supports the hardware and the required data speed. We implemented all the vision architectures with a 100 MHz on-board external clock. Our attempt was to benchmark and compare the performance of different types of vision algorithms on FPGA. Depending on analysis, our objective was to understand the opportunities we have to improve the performance of vision algorithms on modern FPGAs.

The advantages and disadvantages of FPGA-based realization of real-time computer vision applications are also discussed in this section. For the performance evaluation, we used the execution time per single full-scale input image as a measure of the performance.

Furthermore, the utilized resources for the implementation are tabulated and discussed to give an insight view along with limitations of the achievable performance. All the design and implementation methodologies of the architectures analyzed were included in chapter 2.

In our first approach, we developed an image pre-processing architecture in Verilog to analyze the speed of vision algorithms with gate-level parallelism. The results are shown in Fig. 3.1, Fig. 3.2 and Fig. 3.3. They are almost the same as their software-based implementations.

Even though its implementation discarded floating points (Eq. 2.5) and approximations made at the hardware design level (Eq. 2.4), the differences correspond to

27

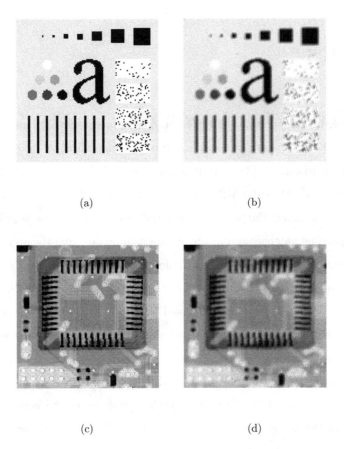

(a) (b)

(c) (d)

Figure 3.1: Results of Gaussian smoothing on two sample images. It can be seen, that smoothing has happened effectively by removing image noise: (a) Original image; (b) Image processed by a 3×3 Gaussian mask; (c) Original image; (d) Image processed by the same Gaussian mask.

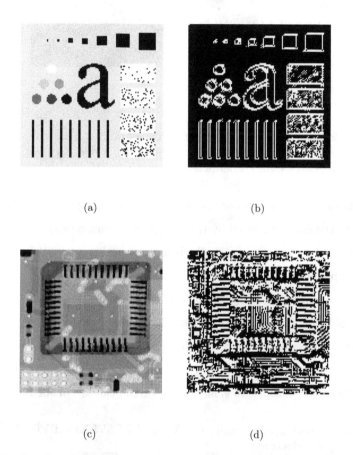

(a)
(b)

(c)
(d)

Figure 3.2: Results of the edge detection on two sample images. The edges are clearly visible and sharpen the image details: (a) Original image; (b) Sobel gradient image; (c) Original image; (d) Image processed by a 3×3 Laplacian mask with $\sigma = 2\sqrt{5}$.

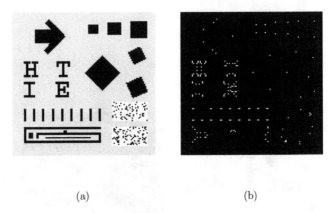

(a)   (b)

Figure 3.3: Results of the right-angle-corner detector. It can identify right-angles in their normal orientations as well as oblique orientations. (a) Original image. (b) Image processed by the corner detector.

pixel values varied at most by one imply that such actions can be readily ignored in return for the speed achieved by this design. Our image pre-processing architecture is capable of operating at 100MHz and processing a 3×3 Sobel operation on 512×512 pixels frame.

The device utilization summary of the image pre-processing architecture is shown in Table 3.1. According to the figures, the resource utilization is very low and there are opportunities to extend the algorithms on top of this architecture. The 86% utilization of block RAM is consumed for image buffering purposes and to extend the algorithms, there is no need to extend the size of the block RAM.

Table 3.1: Device utilization summary on Virtex-5 XC5VLX110T FPGA for image pre-processing architecture.

Resource	Used	Utilization (%)
Slice Registers	747	1
Virtex-5 Slices	522	3
Slice LUTs	1106	1
18Kb BRAMs	256	86

Further, we carried out our experiment for 512×512 images to compare our results with existing systems [18]. The performance comparison is shown in Table 3.2

and Fig. 3.4. The frame transferring delays in the communication core are not considered in the time measuring. In case of evaluating the algorithm's execution speed, it will not make any impact on the results. Results show FPGA implementation is computationally efficient than DSP and GPP. The implementation of our architecture on a 100 MHz XC5VLX110T FPGA is nearly two times faster than the optimized software implementation on a Intel Core 2 Duo @ 2×2.0GHz GPP.

Table 3.2: Timings for a 3×3 Sobel edge detector for a 512×512 image on different platforms and the proposed architecture.

Machine	Architecture	Timing (ms)
Texas TMS320C6414 720MHz [6]	Fixed-point DSP	0.8126
Pentium III Processor 450MHz [4]	GPP	53
Intel Core 2 Duo @ 2×2.0GHz [19]	GPP	1.2530
Proposed FPGA architecture @ 100MHz	FPGA	0.6553

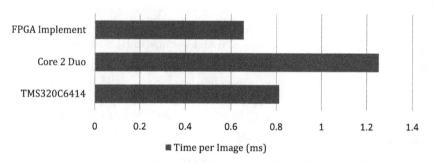

Figure 3.4: Time comparison for a 3x3 Sobel edge detector for a 512x512 image on different platforms and the proposed architecture.

The most difficult thing that we experienced in our first approach was the HDL designing of image processing algorithms is complex and tedious. Therefore, we made our second approach to implement an image coprocessor using HLL synthesis tool. With the use of AccelDSP synthesis tool it efficiently performed floating-to-fixed-point conversion, and hence improved the accuracy of the computations. The effectiveness of the HLL synthesis tool in the prospect of the development speed can be demonstrated by results of image convolution coprocessor, which are shown in Fig. 3.5 and Fig. 3.6. It is a very time consuming process to achieve

31

such an accuracy with floating-point conversion in Verilog implementation. Furthermore, this image coprocessor has the flexibility to data-parallel computation and to increase the size of the convolution kernel. The performance comparison of the image coprocessor with Intel Core 2 Duo processor is given in Table 3.3.

Table 3.3: Speed of the 2-D convolution for 8-bit 800×600 image with different mask sizes

Mask size	FPGA implementation (ms)[1]	Intel Core 2 Duo @ 2×2.0GHz (ms)[2]	Speed ratio
(3×3)	0.57	2.31	4.05
(5×5)	0.96	4.16	4.33
(7×7)	1.34	6.93	5.17
(9×9)	1.72	9.24	5.37
(15×15)	2.88	9.70	3.37

[1] This timing was calculated by extrapolation for maximum resource utilization which does 25 blocks parallelism on Vertex-5 XC5VLX110T FPGA.
[2] OpenCV implementation.

As per the Table 3.3, the implemented architecture in HLS tool is efficient and faster than Intel Core 2 Duo 2×2.0GHz processor. This is mainly because of different parts of the processing which is performed in parallel on the incoming image data. In FPGA, it is essential to implement multi-level parallelism to improve the speed. This is limited by the available resources in the FPGA. Table 3.4 shows the device utilization summary which express resource limitations to parallelism.

Table 3.4: Device utilization summary on Virtex-5 XC5VLX110T FPGA for 2-D convolution with different mask sizes.

Resource	(3×3)	(5×5)	(7×7)	(9×9)	(15×15)
Slice Registers	331 (1%)	333 (1%)	339 (1%)	339 (1%)	345 (1%)
Virtex-5 Slices	260 (1%)	255 (1%)	271 (1%)	251 (1%)	264 (1%)
DSP48Es	3 (4%)	3 (4%)	3 (4%)	3 (4%)	3 (4%)
36Kb BRAMs	128 (86%)	128 (86%)	128 (86%)	128 (86%)	128 (86%)

The primary parallelism inherent in calculation was automatically extracted and pipelined by AccelDSP. The resulting clock frequency, pipeline length, and selected resource utilization of the processing core and its HDL counterpart are shown in Table 3.5. As expected, the HDL version has higher achievable core

Figure 3.5: Results of Gaussian smoothing with different masks: (a) Original image; (b) Result by a 3×3 Gaussian mask ($\sigma = 0.3$); (c) Result by a 5×5 Gaussian mask ($\sigma = 0.6$); (d) Result by a 7×7 Gaussian mask ($\sigma = 1$); (e) Result by a 9×9 Gaussian mask ($\sigma = 1.2$); (f) Result by a 15×15 Gaussian mask ($\sigma = 2.3$).

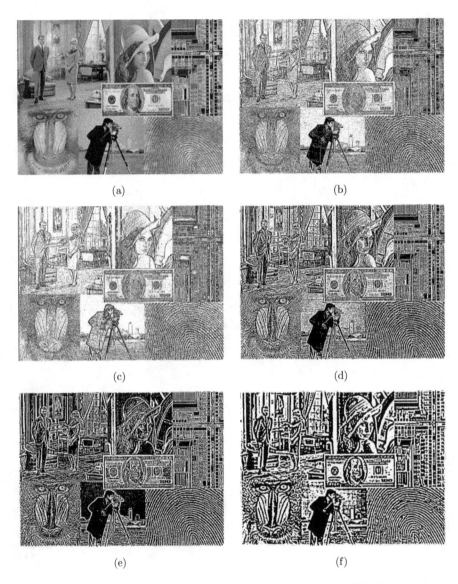

(a)

(b)

(c)

(d)

(e)

(f)

Figure 3.6: Results of Laplace sharpening with different masks: (a) Original image; (b) Result by a 3×3 Laplace mask ($\sigma = 0.3$); (c) Result by a 5×5 Laplace mask ($\sigma = 0.6$); (d) Result by a 7×7 Laplace mask ($\sigma = 1$); (e) Result by a 9×9 Laplace mask ($\sigma = 1.2$); (f) Result by a 15×15 Laplace mask ($\sigma = 2.3$).

frequency because the automated procedure of translating HLL to HDL creates overhead that increases critical-path delays and thus results in a sub-optimal design. The same number of multipliers used in both versions demonstrate that AccelDSP is able to match the efficiency of a hand-coded design with respect to these relatively scarce and expensive resources. Since both designs resulted in similar pipeline length, AccelDSP exhibited potential for exploiting parallelism through pipelining.

Table 3.5: HLS vs. HDL comparisons in image coprocessor design

Core Design	Designing Method	
	HLS	HDL
Core Freq. (MHz)	382	428
Pipeline Len. (cycle)	74	81
MULT18×18s (%)	12	12
Slices (%)	42	40

Next, in our third approach, we implemented processor based architectures to implement software level parallelism on FPGAs to improve performance. We designed our processor based architectures (i.e., PowerPC, single-MicroBlaze and quad-MicroBlaze architectures) by analyzing algorithms into stages and divides them into software threads. The timing results of implemented architectures are as follow.

Table 3.6: Speed of the histogram equalization for (128×128) 8-bit image, implemented with XCL and DDR-RAM.

Architectures	Time Per Frame (s)
Single PowerPC	0.3057
Single MicroBlaze	1.6987
Quad-processor MicroBlaze	0.2500
Intel Core 2 Duo @ 2×2.0GHz	$1.861×10^{-3}$

According to the observed results in Table 3.6, PowerPC architecture is significantly faster than the single Microblaze architecture. However, because of multithreaded computations of quad-Microblaze architecture, it produced similar performance as single PowerPC architecture (Fig. 3.7). As PowerPC processors are not available in all FPGA devices, our quad-Microblaze architecture presents

35

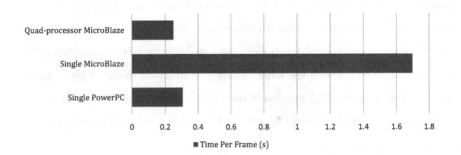

Figure 3.7: Speed of the histogram-equalization for (128×128) 8-bit image, implemented with XCL and DDR-RAM.

an alternative solution in such situations.

Table 3.7: Speed of the histogram equalization for 8-bit image in different resolutions.

Image Resolution	Single MicroBlaze (s)	Quad-processor MicroBlaze (s)	Intel Core 2 Duo @ 2×2.0GHz (s)
(100×100)	1.0631	0.1520	1.034×10^{-4}
(200×200)	4.1849	0.6928	4.988×10^{-4}
(400×400)	16.4967	3.1264	1.779×10^{-3}
(800×800)	67.3725	15.4944	7.189×10^{-3}
(1600×1600)	267.7958	69.2480	2.969×10^{-2}

Furthermore, the speed gain of the quad-microblaze architecture compared to the single-microblaze architecture can be expressed by a performance improvement factor (PIF). It is calculated by Eq. 3.1 where, T_{SM} is time to execute in single-microblaze architecture and T_{QM} is the time to execute in quad-processor-microblaze architecture. The Table 3.9 shows tabulated results of PIF with their respective test case. According to the results, the quad-Microblaze architecture has 75-80% performance improvement compared to its single Microblaze architecture (Fig. 3.8(a)).

$$\text{Performance Improvement Factor (PIF)} = \left(\frac{T_{SM} - T_{QM}}{T_{SM}}\right) \times 100\% \qquad (3.1)$$

Table 3.8: Histogram equalization time per unit image area (100×100).

Image Resolution	Single MicroBlaze (s)	Quad-processor MicroBlaze (s)	Intel Core 2 Duo @ 2×2.0GHz (s)
(100×100)	1.0631	0.1520	1.034×10^{-4}
(200×200)	1.0462	0.1732	1.247×10^{-4}
(400×400)	1.0310	0.1954	1.112×10^{-4}
(800×800)	1.0527	0.2421	1.123×10^{-4}
(1600×1600)	1.0461	0.2705	1.160×10^{-4}

Table 3.9: Performance improvement factor from single-processor to quad-processor for different image resolutions.

Image Resolution	Performance Improvement Factor (%)
(100×100)	85.70
(200×200)	83.45
(400×400)	81.05
(800×800)	77.00
(1600×1600)	74.14

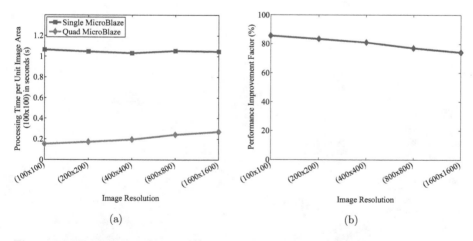

Figure 3.8: Performance improvement comparison from single-Microblaze to quad-Microblazes. (a) Processing time per unit image area (100×100). (b) Performance improvement factor (PIF).

3.1 Summary

According to the benchmark and comparison results, our image pre-processing architecture which could run on a FPGA with 100 MHz clock is nearly two times faster than the optimized software implementation on a Intel Core 2 Duo running at 2×2.0GHz GPP. This image pre-processing architecture consumed fairly low amount of resources. Because of the complexities arose when designing image processing architecture on FPGA in HDL, we also implemented an architecture with HLL synthesis tool. With HLL synthesis tool, we could experience shorter development time in designing image processing architectures. Next, our multi-processor architecture was designed to implement algorithms which are difficult to gain speed by hardware parallelism. In particular, our quad-Microblaze architecture could perform histogram equalization 75-80% efficiently compared to its single Microblaze architecture. However this quad-Microblaze implementation is considerably slower than DSPs/GPPs.

CHAPTER 4

Conclusion

In this research we implemented several low-level vision algorithms on FPGA and evaluated their performance. Based on the results of the evaluations, FPGA implementations outperform the DSP and GPP implementations for algorithms which can exploit a large number of parallelism. In particular, low-level algorithms such as convolution which can run in parallel fit best to FPGAs. However, the processing speed for sequential computations (e.g., algorithms on PowerPC/MicroBlaze) is slower than on DSPs/GPPs because of the higher clock frequency of DSPs/GPPs.

With the use of HLL synthesis tools (e.g., Xilinx AccelDSP), it is possible to design FPGA-based vision architectures in a short development time. HLL synthesis tools include automated and flexible floating-to-fixed-point conversion, which is inefficient and time-consuming to approach manually. However, the common drawback of HLL synthesis tools is their requirement for a specific programming style to allow efficient implementation of parallelism and generation of optimized designs. Furthermore, most of these tools are vendor-specific.

In the research we also worked to benchmark the performance of multi-processor architecture on FPGA. Our quad-Microblaze architecture is suitable to implement algorithms which are harder to exploit parallelism in hardware level. However, it needs to consider critical design aspects related to memory access, data bandwidth and bus protocol overhead, which are time consuming tasks. In some instances, the bus protocol overhead is comparatively large as compared to the actual exe-

39

cution time. This consumes away any speed advantage gained trough hardware acceleration. Finally, this bus connection might be a data bandwidth bottleneck for some image processing tasks.

4.1 Future Directions

The future direction of this research will be focused on investigating how key algorithms in computer vision related to feature detection, segmentation and tracking can be implemented by using our developed architectures on FPGA. The main objective will be implementing a system to achieve above mentioned vision tasks with real-world image sequences, while benchmarking their performance.

References

[1] "Texas instruments - video and vision guide," pp. 29–32, December 2009.

[2] "Accelerating high-performance computing with FPGAs," *Altera FPGA solutions - White Paper*, pp. 1–5, January 2010.

[3] L. B. W. Jensen, A. Kjaer-Nielsen, J. D. Alonso, E. Ros, and N. Kruger, "A hybrid FPGA/coarse parallel processing architecture for multi-modal visual feature descriptors," *International Conference on Reconfigurable Computing and FPGAs*, pp. 241–246, December 2008.

[4] C. Torres-Huitzil and M. Arias-Estrada, "An FPGA architecture for high speed edge and corner detection," *Fifth IEEE International Workshop on Computer Architectures for Machine Perception*, pp. 112–116, 2000.

[5] Z. Wei, D. J. Lee, and B. E. Nelson, "FPGA-based real-time optical flow algorithm design and implementation," *Journal of Multimedia on Academy Publisher*, vol. 2, no. 5, pp. 38–45, September 2007.

[6] D. Baumgartner, P. Rossler, and W. Kubinger, "Performance benchmark of DSP and FPGA implementations of low-level vision algorithms," *IEEE Conference on Computer Vision and Pattern Recognition*, pp. 1–8, June 2007.

[7] J. Teifel and R. Manohar, "Highly pipelined asynchronous FPGAs," *Proceedings of the ACM/SIGDA 12th International Symposium on Field Programmable Gate Arrays*, pp. 133–142, February 2004.

[8] V. Asokan, "Designing multiprocessor systems in platform studio," *White Paper for Xilinx Platform Studio (XPS)*, vol. 2, pp. 1–4, November 2007.

[9] K. T. Gribbon, D. G. Bailey, and C. T. Johnston, "Using design patterns to overcome image processing constraints on FPGAs," *Third IEEE International Workshop on Electronic Design, Test and Applications (DELTA 2006)*, pp. 47–56, January 2006.

[10] J. Curreri, S. Koehler, A. D. George, B. Holland, and R. Garcia, "Performance analysis framework for high-level language applications in reconfigurable computing," *ACM Transactions on Reconfigurable Technology and Systems*, vol. 3, no. 1, 2010.

[11] J. L. Tripp, M. Gokhale, and K. D. Peterson, "Trident: From high-level language to hardware circuitry," *IEEE Computer*, vol. 40, no. 3, pp. 28–37, 2007.

[12] S. Y. C. Li, G. C. K. Cheuk, K.-H. Lee, and P. H. W. Leong, "FPGA-based SIMD processor," *11th IEEE Symposium on Field-Programmable Custom Computing Machines (FCCM 2003)*, pp. 267–268, April 2003.

[13] S. McBader and P. Lee, "An FPGA implementation of a flexible, parallel image processing architecture suitable for embedded vision systems," *Parallel and Distributed Processing Symposium*, vol. International Volume, pp. 22–26, April 2003.

[14] "Single-port block memory core v6.2," *Xilinx LogiCORE*, April 2005.

[15] M. Samarawickrama, A. Pasqual, and R. Rodrigo, "FPGA-based compact and flexible architecture for real-time embedded vision systems," *Fourth IEEE International Conference on Industrial and Information Systems*, pp. 337–342, December 2009.

[16] K. Lindberg and K. Nissbrandt, "An evaluation of methods for FPGA implementation from a matlab description," *Masters Degree Thesis, Stockholm, Sweden*, 2008.

[17] G. Ottoni, R. Rangan, A. Stoler, M. J. Bridges, and D. I. August, "From sequential programs to concurrent threads," *IEEE Computer Architecture Letters*, vol. 5, no. 1, pp. 6–9, 2006.

[18] N. K. Ratha and A. K. Jain, "FPGA-based computing in computer vision," *Fourth IEEE International Workshop on Computer Architecture for Machine Perception*, pp. 128–137, October 1997.

[19] B. Kisacanin, S. S. Bhattacharyya, and S. Chai, *Embedded Computer Vision*. Springer Publishing Company, 2009.

www.ingramcontent.com/pod-product-compliance
Lightning Source LLC
Chambersburg PA
CBHW060442060326
40690CB00019B/4306